Abdelhafid Mimouni

Cellular Signaling: A Bioinorganic Exploration

Abdelhafid Mimouni

Cellular Signaling: A Bioinorganic Exploration

ScienciaScripts

Imprint

Any brand names and product names mentioned in this book are subject to trademark, brand or patent protection and are trademarks or registered trademarks of their respective holders. The use of brand names, product names, common names, trade names, product descriptions etc. even without a particular marking in this work is in no way to be construed to mean that such names may be regarded as unrestricted in respect of trademark and brand protection legislation and could thus be used by anyone.

Cover image: www.ingimage.com

This book is a translation from the original published under ISBN 978-620-6-72099-7.

Publisher:
Sciencia Scripts
is a trademark of
Dodo Books Indian Ocean Ltd. and OmniScriptum S.R.L publishing group

120 High Road, East Finchley, London, N2 9ED, United Kingdom
Str. Armeneasca 28/1, office 1, Chisinau MD-2012, Republic of Moldova, Europe
Printed at: see last page
ISBN: 978-620-8-05743-5

CELLULAR SIGNALLING: A BIOINORGANIC EXPLORATION

AUTHOR

Dr. Abdelhafid Mimouni

An independent researcher in bioinorganic chemistry, Dr Mimouni is an expert in macromolecular synthesis and characterisation. He obtained his PhD in Chemistry from the University of Paris XII in 1997, after a Diplôme des Études Approfondies in Bioinorganic Systems from the University of Paris XI in 1993, where he also obtained his Licence and Maîtrise in Chemistry.

SUMMARY

His book examines cell signalling and the importance of metal ions such as calcium, zinc and magnesium in this process. It explains how these ions act as internal messengers and how their imbalance can lead to cellular dysfunction. The book also details techniques for analysing these mechanisms and explores the role of glutathione in cellular detoxification. Finally, it offers perspectives on the future applications of this knowledge for advanced scientific interventions.

INDEX

INTRODUCTION

Cell signalling is a fundamental process that enables cells to communicate and respond to their environment. This complex mechanism is essential to the proper functioning of living organisms, regulating functions such as growth, metabolism, immune response and neuronal communication. Among the many players involved, metal ions play a crucial role that is often underestimated.

Metal ions such as calcium (Ca^{2+}), zinc (Zn^{2+}) and magnesium (Mg^{2+}) are not only building blocks of biomolecules, but also dynamic regulators of biological processes. Their concentration, distribution and interaction with other molecules profoundly influence signalling pathways. Calcium, for example, is known for its role as a second messenger, while zinc is involved in regulating gene expression and modulating cellular responses.

The aim of this book is to explore in depth the role of metal ions in cell signalling, highlighting the mechanisms by which they influence signalling pathways and cell regulation. We will examine the interactions between these ions and various receptors, as well as the implications of ion dysregulation in the development of pathologies. Through this exploration, we aim to provide an enriched understanding of the importance of metal ions in biological processes, while highlighting potential therapeutic perspectives. This book is aimed at researchers, students and professionals interested in bioinorganics and cell biology, providing them with essential knowledge on a rapidly expanding field rich in discoveries.

CHAPTER 1
FUNDAMENTALS OF CELL SIGNALLING

Definition of cell signalling

Cell signalling is the process by which cells communicate with each other and respond to their environment. It involves the transmission of information through chemical or physical signals, enabling cells to coordinate their functions, develop, differentiate and maintain homeostasis. This mechanism is crucial to the proper functioning of multicellular organisms, regulating processes such as growth, immune response and hormone regulation.

Signal Types

1. **Chemical signals**

o **Hormones**: Produced by endocrine glands, they circulate in the blood and influence various bodily functions (for example, insulin regulates glucose metabolism).

o **Neurotransmitters**: Released by neurons, they transmit signals between nerve cells (e.g. dopamine, serotonin).

o **Cytokines**: Proteins involved in communication between immune system cells, playing a key role in inflammatory and immune responses.

2. Physical signals

o **Light**: Photoreceptors react to light, influencing processes such as photosynthesis in plants and vision in animals.

o **Pressure and temperature**: Cells detect changes in pressure or temperature, which are essential for physiological responses such as pain or thermoregulation.

Signalling channels Main

1. Cyclic AMP (cAMP) pathway

o Involved in the regulation of numerous cellular functions, this pathway is activated by hormones such as adrenaline. cAMP, as a second messenger, amplifies the signal and leads to various cellular responses.

2. MAP kinase pathway

o Crucial for cell proliferation, differentiation and survival, this pathway is activated by growth factors and involves a series of phosphorylations that modify the activity of target proteins.

3. PI3K/Akt pathway

o Involved in cell survival and metabolism, this pathway is often activated by growth signals and contributes to the regulation of cell growth and resistance to apoptosis.

4. Nuclear receptor signalling pathway

o These receptors, which bind fat-soluble hormones such as steroids, act directly on DNA to regulate gene expression.

CHAPTER 2
METAL IONS AS SECOND MESSENGERS

Definition of second messengers

Second messengers are intracellular molecules that play an essential role in transmitting and amplifying signals received by membrane receptors. Unlike first messengers, such as hormones or neurotransmitters, which bind to receptors on the cell surface, second messengers act inside the cell to trigger cascades of biochemical reactions. In this way, they enable a rapid, coordinated response to various external stimuli. Second messengers include ions, cyclic nucleotides and lipids, each with specific mechanisms of action and regulation.

Role of metal ions (Ca^{2+}, Zn^{2+}, Mg^{2+}) in signalling

Calcium (Ca^{2+}) Calcium is one of the most studied and crucial second messengers in cell signalling. It influences various physiological processes, including :

- **Muscle contraction**: An increase in intracellular Ca^{2+} concentration triggers muscle contraction by interacting with proteins such as troponin.
- **Release of neurotransmitters**: In neurons, an action potential causes calcium channels to open, allowing rapid entry of Ca^{2+}, leading to the release of neurotransmitters at the synapse.

- **Enzyme activation**: Ca^{2+} activates various enzymes, such as kinases, thereby modifying the function of target proteins.

Zinc (Zn^{2+}) Zinc, although less studied than calcium, also plays an important role as a second messenger:

- **Regulation of gene expression**: Zn^{2+} can modulate the activity of transcription factors, influencing the expression of specific genes.
- **Protection against oxidative stress**: As a cofactor for antioxidant enzymes, zinc helps protect cells against oxidative damage, influencing signalling pathways linked to cell survival.

Magnesium (Mg^{2+}) Magnesium, often considered a support element, also plays a role in cell signalling:

- **Enzyme cofactor**: Mg^{2+} is essential for the activity of many enzymes, particularly those involved in protein phosphorylation, influencing signalling pathways based on phosphorylation.
- **Membrane stabilisation**: By stabilising cell membranes, magnesium helps to maintain cell integrity and regulate the passage of ions, which can affect signalling processes.

Activation and regulation mechanisms

The mechanisms by which metal ions are activated and regulated as second messengers are complex and involve several steps:

- **Ion release** : Various stimuli, such as activation of membrane receptors or changes in membrane potential, can trigger the release of metal ions into the cell.

- **Ion channels**: Specific channels, such as voltage-gated calcium channels, allow rapid entry of Ca^{2+} in response to electrical signals.

- **Transporters and pumps** : Membrane transporters regulate ion levels by controlling their entry into and exit from the cell.

- **Signal amplification**: Once released, metal ions can activate target proteins, leading to a cascade of reactions that amplify the initial signal. For example, a small increase in Ca^{2+} can lead to massive activation of kinases, amplifying the cellular response.

- **Signal regulation**: Metal ion levels are finely regulated by feedback mechanisms. Binding proteins, pumps and exchangers adjust ion concentrations to maintain cellular equilibrium.

- **Integrated signalling**: Metal ions often interact with other signalling pathways. For example, Ca^{2+} can act in synergy with with messengers such as cyclic AMP to coordinate complex cellular responses.

CHAPTER 3
METAL IONS AND RECEPTORS

Interaction of metal ions with membrane receptors

Metal ions play a crucial role in modulating the activity of membrane receptors, thereby influencing the transmission of cellular signals. Their interaction with these receptors is essential for cell signalling, as it can determine the response of cells to various external stimuli. Metal ions such as calcium (Ca^{2+}), zinc (Zn^{2+}) and magnesium (Mg^{2+}) interact with different types of receptors, in particular ionotropic and metabotropic receptors. These interactions can be direct, by binding to specific sites on the receptors, or indirect, by modulating associated signalling pathways.

Ionotropic and metabotropic receptors :

• **Ionotropic receptors** : These receptors function as ion channels, directly regulating the passage of ions across the cell membrane. Metal ions can modulate their opening and closing, thereby influencing the cellular response.

• **Metabotropic receptors** : These receptors are coupled to G proteins and influence intracellular signalling pathways via second messengers. Metal ions can modulate the activity of these receptors by affecting the resulting signalling cascades.

Examples of ion-sensitive receptors Neurotransmitter receptors :

• **NMDA (N-methyl-D-aspartate) receptors**: These glutamatergic receptors are sensitive to calcium. Activation of NMDA receptors by glutamate facilitates Ca^{2+} entry into the cell, which is crucial for synaptic plasticity and memory formation. Ca^{2+} levels also regulate the sensitivity of NMDA receptors, influencing cognitive and neuronal processes.

• **GABA (gamma-aminobutyric acid) receptors** : Some GABA_A receptors are sensitive to Zn^{2+} concentration. Zinc modifies the activity of these receptors, impacting inhibitory transmission in the central nervous system and playing a role in the brain's excitatory/inhibitory balance.

Hormone receptors :

• **Calcitonin receptors**: These receptors are activated by the hormone calcitonin, which regulates blood calcium levels. The interaction of calcitonin with its receptor triggers a cascade of intracellular signals that reduce the release of Ca^{2+} from bone, thereby helping to regulate calcium metabolism and bone homeostasis.

• **Glucocorticoid receptors**: These receptors, activated by steroid hormones, are modulated by Mg^{2+}. Magnesium plays a role in stabilising cell membranes and can influence the response of cells to glucocorticoids, affecting processes such as inflammation, metabolism and the stress response.

Impact on intracellular signalling

The interaction of metal ions with membrane receptors has significant effects on intracellular signalling, with several key impacts:

• **Signal amplification**: The entry of ions such as Ca^{2+} into the cell can trigger cascades of biochemical reactions that amplify the initial signal. For example, activation of NMDA receptors induces a large influx of Ca^{2+}, leading to the activation of kinases and other proteins that modulate various cellular functions, such as growth and survival.

• **Modulation of cellular response**: Metal ions can influence the sensitivity of receptors to their ligands. High concentrations of Zn^{2+}, for example, can inhibit the activity of NMDA receptors, altering the neuronal response to stimuli and playing a role in processes such as synaptic plasticity and neuropsychiatric disorders.

• **Signal integration**: Metal ions enable cells to integrate signals from different signalling pathways. Ca^{2+}, for example, can act in synergy with other second messengers, such as cyclic AMP, to coordinate complex cellular responses such as muscle contraction, hormone secretion and metabolic regulation.

• **Regulation of gene expression**: Variations in metal ion concentrations can modulate gene expression by activating specific transcription factors. Zn^{2+}, in particular, can influence the activity of regulatory proteins, affecting the transcription of genes involved in processes such as cell growth, differentiation and stress response.

CHAPTER 4
METALS AND SPECIFIC SIGNALLING PATHWAYS

Calcium-related signalling pathways

Calcium (Ca^{2+}) plays a crucial role as a second messenger in many cell signalling pathways. Its intracellular concentration is regulated with remarkable precision by pumps, exchangers and ion channels. When an external stimulus is perceived, a rapid increase in intracellular Ca^{2+} can occur, triggering a series of complex biological events.

Role in muscle contraction

In muscle cells, calcium is fundamental to the contraction process. The arrival of an action potential at the muscle cell membrane causes voltage-dependent calcium channels to open, allowing Ca^{2+} to enter the cell. This calcium binds to troponin, a regulatory protein that facilitates interaction between actin and myosin, the main contractile proteins in muscles. This mechanism is essential not only for skeletal muscle, but also for cardiac muscle and smooth muscle, regulating various bodily functions such as heart contraction and intestinal peristalsis.

Role in neurotransmission

In the nervous system, Ca^{2+} is essential for neurotransmission. When neurons are activated, the entry of Ca^{2+} through voltage-dependent calcium channels leads to the release of neurotransmitters at the synapse.These neurotransmitters bind to receptors on postsynaptic neurons, facilitating the transmission of the nerve signal. Modulation of Ca^{2+} concentration can influence synaptic strength and neuronal plasticity, thus affecting complex cognitive processes such as learning and memory. Alterations in Ca^{2+} regulation may be associated with neurological disorders such as Alzheimer's disease and depression.

The role of zinc in cell signalling

Zinc (Zn^{2+}) is a versatile metal ion that plays a regulatory role in several cellular processes. It is involved in cell signalling in a variety of ways and influences many aspects of cell function.

Effects on gene transcription

Zn^{2+} is essential for the function of many transcription factors. By binding to specific sites, it modifies the activity of regulatory proteins and influences the expression of target genes. For example, zinc can activate transcription factors such as transcription factor 1 (TF1), which regulates the expression of genes involved in the stress response and inflammation. In addition, Zn^{2+} is crucial for the structure of certain proteins, stabilising zinc finger motifs that are fundamental for DNA recognition and regulation, thus influencing the transcription of genes linked to cell growth and the stress response.

Immune response

Zinc plays a central role in the immune response. It is necessary for the development and function of T and B lymphocytes, two essential types of immune cell. Zinc deficiency can compromise the immune response, increasing susceptibility to infection and slowing wound healing. In addition, Zn^{2+} regulates the production of cytokines, signalling molecules that orchestrate the immune response, influencing communication between immune cells and the coordination of the inflammatory response.

Involvement of magnesium in signalling

Magnesium (Mg^{2+}) is another essential metal ion involved in many signalling pathways. It acts primarily as a cofactor for many enzymes and plays a role in stabilising cell membranes.

Role in cell signalling

Mg^{2+} is involved in the regulation of calcium signalling. It helps maintain ion balance in cells and can modulate Ca^{2+} entry through calcium channels, regulating their opening and closing. An adequate concentration of Mg^{2+} is necessary for the optimal function of the ion channels and pumps that regulate Ca^{2+} levels, ensuring the smooth functioning of cellular signalling pathways. From Imbalances in Mg^{2+}

levels can disrupt these processes, affecting muscle contraction, neurotransmission and other essential cellular functions.

Effects on health

Insufficient levels of magnesium can lead to dysfunctions in cell signalling, contributing to cardiovascular disease, neurological disorders and metabolic problems. Mg^{2+} is also involved in modulating the inflammatory response, acting as a regulator to limit excess inflammation and prevent chronic inflammatory pathologies. Research shows that magnesium plays a protective role against a variety of conditions, including metabolic disorders such as type 2 diabetes and cardiovascular disease.

CHAPTER 5
DYSREGULATION AND PATHOLOGIES

Consequences of metal ion dysregulation

Metal ions such as calcium (Ca^{2+}), zinc (Zn^{2+}) and magnesium (Mg^{2+}) play essential roles in various physiological processes, including cell signalling, muscle contraction and regulation of the nervous and immune systems. An imbalance in their concentrations can have serious repercussions on cellular and organ health, leading to a variety of pathologies.

Calcium (Ca^{2+})

Hypercalcaemia, or excess intracellular Ca^{2+}, can cause significant damage to cells, including death by apoptosis or necrosis. This phenomenon is often associated with neurodegenerative disorders and cardiovascular pathologies. Conversely, Ca^{2+} deficiency can disrupt muscle contraction, alter neurotransmission and lead to blood clotting problems, increasing the risk of bone fractures and heart disease.

Zinc (Zn^{2+})

Zinc imbalance is also a cause for concern. Zn^{2+} deficiency is linked to reduced immune function, impaired wound healing and cognitive impairment. On the other hand, too much zinc can be neurotoxic, disrupting cell signalling and having deleterious effects on the immune system. It can also have an adverse effect on the nervous system, in particular by promoting memory problems and cognitive deficits.

Magnesium (Mg^{2+})

Magnesium is crucial for numerous enzymatic reactions and the regulation of neuronal excitability. Mg^{2+} deficiency is associated with metabolic disorders, muscle cramps and cardiac arrhythmias. Although less common, hypermagnesaemia can also lead to adverse effects, such as central nervous system depression, affecting coordination and cognitive function.

Links between ion imbalance and diseases Neurodegenerative diseases

Ionic imbalances, particularly those involving Ca^{2+} and Zn^{2+}, are closely linked to neurodegenerative diseases such as Alzheimer's and Parkinson's.

- **Alzheimer's disease**: Excessive accumulation of Ca^{2+} in neurons can cause excitotoxicity, leading to neuronal death. At the same time, high levels of Zn^{2+} can disrupt the function of amyloid proteins, aggravating the formation of senile plaques, one of the markers of the disease.

- **Parkinson's disease**: dysregulation of Ca^{2+} and Zn^{2+} levels is associated with oxidative stress and inflammation, two key factors in the degeneration of dopaminergic neurons, contributing to the motor symptoms of the disease.

Cardiovascular diseases

Ionic imbalances, particularly those of Mg^{2+} and Ca^{2+}, are also implicated in cardiovascular disease.

- **Hypertension**: A deficiency in Mg^{2+} is often correlated with an increase in blood pressure, as magnesium plays a role in relaxing blood vessels and regulating blood pressure.
- **Cardiac arrhythmias**: Abnormal Ca^{2+} levels can disrupt cardiac excitability, leading to potentially serious arrhythmias and heart rhythm disorders.

Case studies

Case study 1: Alzheimer's disease

Research has shown that Alzheimer's patients have abnormal levels of Ca^{2+} and Zn^{2+} in their cerebrospinal fluid. Interventions aimed at regulating these ions, such as the use of calcium or zinc chelators, have shown results. have shown promise in improving cognitive symptoms in some patients, underlining the importance of ion regulation in the management of this disease.

Case study 2: Metabolic syndrome

Studies have shown that individuals suffering from metabolic syndrome are often deficient in Mg^{2+}, contributing to complications such as hypertension and type 2 diabetes. Magnesium supplementation has shown significant beneficial effects on blood sugar regulation and blood pressure, suggesting that magnesium plays a protective role against these metabolic disorders.

Case study 3: Parkinson's disease

A clinical study observed that patients with Parkinson's disease have elevated levels of Ca^{2+} in dopaminergic neurons, which is associated with increased neuronal degeneration. Treatments aimed at reducing excess Ca^{2+}, such as the use of calcium channel blockers, have shown significant improvements in motor function in a subset of patients, highlighting the potential impact of ion regulation in managing the symptoms of the disease.

CHAPTER 6
EXPERIMENTAL APPROACHES AND STUDY TECHNIQUES

Methods for analysing metal ions in cells

Understanding the role of metal ions in cell signalling and associated pathologies relies on precise analysis techniques. Here is an overview of the main methods used to quantify and localise these ions in cells:

Fluorescence spectroscopy

Fluorescence spectroscopy is widely used to detect metal ions in cells using specific probes. Calcium indicators such as Fura-2 or Fluo-4, for example, allow real-time visualisation of variations in ionic concentration. By binding to the targeted ions, these probes emit a fluorescence whose intensity varies according to the ion concentration, providing dynamic information on ionic changes at the cellular level.

Mass spectrometry

Mass spectrometry (MS) is a highly accurate technique for quantifying metal ions in complex biological samples. By analysing biological tissues or fluids, MS provides detailed data on the concentration and identity of the ions present. This method is essential for studies requiring high quantitative precision and in-depth analysis of ionic composition.

X-ray microanalysis

X-ray microanalysis, such as X-ray fluorescence (EDX), provides detailed images of the spatial distribution of metal ions in cells. This technique is particularly useful for examining the subcellular location of ions and their relationship with other cellular structures, providing crucial information about ion distribution at the microscopic level.

Signalling measurement techniques

Several techniques for measuring signalling are used to assess cellular responses to stimuli:

Imaging

Cellular imaging, in particular fluorescence microscopy and electron microscopy, makes it possible to observe changes in the localisation and dynamics of metal ions within cells. Super-resolution microscopy, such as STED (Stimulated Emission Depletion), provides nanometric resolutions enabling detailed study of ionic interactions at a very fine level, providing an in-depth understanding of the underlying ionic mechanisms.

Biochemical techniques

Biochemical techniques, such as enzyme assays and colorimetric assays, measure enzymatic activities influenced by metal ions. For example, calcium release assays can be used to assess the response of cells to various stimuli, providing information on the impact of metal ions on cellular biochemical processes.

Electrophysiology

Electrophysiology is a crucial method for studying the effects of metal ions on cellular excitability. Action potential recordings and ion currents allow us to measure how ions influence electrical signalling in neurons and muscle cells. This technique helps to understand the effects of ions on the propagation of electrical signals and neuronal communication.

In vitro and in vivo studies

In vitro and in vivo studies play a complementary role in understanding the role of metal ions in cell signalling.

In vitro studies

In vitro studies are carried out on cell cultures or isolated tissues, allowing precise control of experimental conditions and examination of the effects of metal ions in a simplified environment. For example, by exposing neuron cultures to varying concentrations of Ca^{2+} or Zn^{2+}, it is possible to observe the impact on neurotransmission and cell survival in a controlled setting.

In vivo studies

In vivo studies, using animal models, make it possible to examine the effects of metal ions in an entire organism. These studies assess the complex interactions between ions, tissues and biological systems. In vivo imaging techniques, such as functional MRI and positron emission tomography (PET), are used to visualise ionic changes in real physiological contexts, offering insights into the overall biological effects of metal ions.

Combined approaches

Integrating data from in vitro and in vivo studies is essential for validating experimental results and understanding the underlying mechanisms of ion dysregulation. Results obtained in cell cultures can be confirmed by in vivo studies, establishing more robust links between ionic imbalances and specific pathologies. This combined approach provides a comprehensive and in-depth view of ionic processes at both cellular and systemic levels.

CHAPTER 7
CELLULAR DETOXIFICATION VIA GLUTATHIONE

Detoxification is a crucial process for maintaining cellular balance and preventing damage from toxins and excess metal ions. Glutathione, a ubiquitous tripeptide, plays a central role in these detoxification mechanisms, facilitating the elimination of harmful substances and protecting cells against oxidative effects. This chapter explores in detail the role of glutathione in detoxification, highlighting its mechanisms and interactions with other cellular processes.

1. The role of Glutathione in detoxification

Glutathione (GSH) is a powerful antioxidant that plays a multifunctional role in cell detoxification. Its main roles are as follows:

• **Metal ion chelation**: Glutathione binds to metal ions to form soluble complexes. This chelation reduces the toxicity of metal ions by facilitating their elimination via excretory pathways. Glutathione-metal complexes are less reactive and less likely to cause oxidative damage.

• **Neutralisation of Reactive Oxygen Species (ROS)**: As an antioxidant, glutathione neutralises the ROS generated during the metabolism of excess metal ions. This action reduces oxidative stress and protects cellular structures such as lipids, proteins and nucleic acids.

- **Facilitates excretion**: Glutathione helps to export excess metal ions from cells by forming soluble complexes. These complexes are then transported to excretory organs such as the liver and kidneys for elimination from the body.

- **Support for other detoxification mechanisms**: Glutathione works in synergy with other detoxification systems, such as metallothionins. By reducing the load on these proteins, glutathione optimises their effectiveness in managing metal ions.

2. Glutathione Regulation Mechanisms

Regulation of glutathione levels is essential for effective detoxification and protection against oxidative damage. The main mechanisms include :

- **Glutathione synthesis**: Glutathione is synthesised from its precursors, cysteine, glutamic acid and glycine. Glutamyl cysteine ligase, a key enzyme, catalyses the first stage of this synthesis, while glutathione synthetase completes the process.

- **Transcriptional regulation**: Transcription factors such as Nrf2 (Nuclear factor erythroid 2-related factor 2) regulate the expression of genes involved in glutathione synthesis and regulation. Nrf2 responds to oxidative stress signals by increasing glutathione production.

- **Control of redox metabolism**: Glutathione exists mainly in two forms: reduced (GSH) and oxidised (GSSG). The balance between these two forms is regulated by cellular mechanisms, influencing glutathione's antioxidant capacity.

- **Preservation against Degradation**: Cellular mechanisms which limit the degradation of glutathione, including inhibitors of degradation enzymes, help to maintain optimal levels of this tripeptide.

3. Interactions of Glutathione with Metal Ions

Glutathione interacts in a complex way with various metal ions, influencing their detoxification and regulation:

- **Calcium**: Glutathione influences the regulation of intracellular calcium, which in turn affects processes such as muscle contraction and neuronal signalling. Adequate calcium balance is crucial for cell function.
- **Zinc**: Zinc is essential for the function of antioxidant enzymes, and glutathione helps to modulate zinc levels by forming soluble complexes, thereby reducing its oxidative impact.
- **Iron**: When iron is in excess, it can generate ROS. Glutathione helps to chelate iron, limiting its capacity to induce oxidative damage.

4. Impact on Cellular Signalling

Glutathione and metal ion levels have a profound impact on cell signalling:

- **Redox signalling pathways**: Glutathione regulates redox signalling pathways, influencing cellular responses to oxidative stress. Altered glutathione levels can disrupt these pathways, leading to cellular dysfunction.

- **Transcription factors**: Glutathione and metal ions affect the activity of transcription factors, modifying gene expression and the cellular response to external stimuli.

- **Responses to Environmental Stress**: Cells adjust their glutathione levels in response to environmental stresses, such as toxins and pathogens. This process enables cells to survive and maintain their functional integrity.

5. Associated disturbances and pathologies

Imbalances in glutathione and metal ion levels can be associated with various pathologies:

- **Neurological diseases**: Disturbances in the regulation of glutathione and metal ions are linked to neurological diseases, where they affect neuronal signalling and brain function.

- **Cardiovascular disease**: Calcium and glutathione imbalances can contribute to cardiovascular disease by altering cardiac signalling and promoting oxidative states.

Conclusion

Glutathione plays a crucial role in cellular detoxification, by chelating metal ions, neutralising ROS and facilitating the elimination of toxins. Its regulation is essential for maintaining effective detoxification and protecting cells against oxidative damage. The complex interactions between glutathione and metal ions influence many aspects of cell biology, from signalling to associated pathologies.

CHAPTER 8
THERAPEUTIC PERSPECTIVES

Potential targets for treatments based on metal ions

Because of their crucial role in various biological processes, metal ions represent interesting targets for the development of therapeutic treatments. Here is an overview of the main targets and associated therapeutic approaches:

Calcium (Ca^{2+})

Calcium plays an essential role in muscle contraction, neurotransmission and cell signalling. Drugs targeting calcium channels, such as calcium channel blockers, are already being used to treat conditions such as hypertension and heart disease. By modulating the entry of Ca^{2+} into cells, these drugs help to regulate muscle contraction and improve heart function.

Zinc (Zn^{2+})

Zinc is involved in immune function, the regulation of gene expression and various enzymatic processes. Therapies aimed at modulating zinc levels could be developed to treat neurodegenerative diseases, such as Alzheimer's, and immune disorders. Zinc chelating agents, which eliminate excess zinc from the body, are being considered to treat pathological conditions associated with zinc overload.

Magnesium (Mg^{2+})

Magnesium is crucial for numerous enzymatic reactions and for regulating neuronal excitability. Magnesium supplements are being explored as potential treatments for conditions such as migraines, depressive disorders and cardiovascular disease. Magnesium plays a role in modulating the inflammatory response and in regulating blood pressure, making it relevant to a variety of pathologies.

Development of drugs modulating ion signalling

The development of drugs targeting ion signalling offers promising prospects for the treatment of various pathologies. Here are a few current approaches:

Ion channel modulators

Ion channel modulators, including agonists and antagonists, influence cell signalling by regulating the entry and exit of ions. For example, NMDA receptor antagonists, which modulate Ca^{2+} entry, are being explored for the treatment of neurodegenerative diseases such as Alzheimer's disease. These modulators can help to balance ion levels and prevent the harmful effects of over-stimulation.

Ion chelators

Ion chelators, which bind to metal ions to facilitate their elimination from the body, are already used to treat iron and copper overload. Research is underway to develop specific chelators for zinc and manganese. These agents could help to treat pathological conditions associated with excesses of these ions, offering a targeted approach to reducing their toxicity.

Gene therapies

Gene therapies represent an innovative avenue for targeting the mechanisms that regulate metal ions. By using gene vectors to introduce genes encoding proteins involved in ion transport or regulation, it is possible to modulate the expression of regulatory proteins. This approach could offer targeted solutions for specific diseases by directly influencing the levels and distribution of metal ions.

Applications in regenerative medicine and oncology

Metal ions and their modulation also have promising applications in regenerative medicine and oncology:

Regenerative medicine

In the field of regenerative medicine, metal ions such as calcium and magnesium play a role in cell differentiation and tissue regeneration. For example, appropriate levels of Ca^{2+} can promote the differentiation of stem cells into bone cells, which is exploited for bone repair therapies. Approaches using metal ions to modulate cell growth and regeneration are being developed to improve treatments for injuries and degenerative diseases.

Oncology

Metal ions are involved in tumour progression and can be targeted to inhibit cell proliferation and induce apoptosis in cancer cells. For example, treatments aimed at modifying the concentration of Zn^{2+} could influence tumour growth by affecting specific signalling pathways. Modulating ion signalling in cancer cells offers a potential approach for improving therapeutic strategies against cancer.

Nanomedicine

Nanomedicine, which uses nanoparticles to deliver drugs in a targeted manner, is also exploring the properties of metal ions. Nanoparticles containing ions such as iron or gold are being studied for their ability to specifically target tumour cells while minimising side-effects on healthy cells. This approach offers interesting prospects for more effective and less invasive treatments.

CONCLUSION

In conclusion, this book has explored in depth the crucial role of metal ions in cell signalling and their impact on cellular function. We have seen how ions such as calcium, zinc and magnesium act as internal messengers, modulating essential processes and influencing various signalling pathways. The role of glutathione in detoxification and its interactions with these ions add an important dimension to our understanding of cellular mechanisms. The experimental and technical approaches discussed provide valuable tools for future research, while the therapeutic perspectives outlined highlight the potential application of this knowledge in a variety of fields. This work highlights the importance of these elements in cell biology and paves the way for new explorations in scientific research.

LEXICON

• **Cell signalling**: Process by which cells communicate and react to internal or external signals.

• **Hormone**: Molecule produced by endocrine glands which regulates physiological functions.

• **Neurotransmitter**: Chemical substance that transmits a signal between neurons.

• **Cytokine**: Protein involved in communication between immune cells.

• **Second messenger**: intracellular molecule that transmits a signal from a membrane receptor to other targets in the cell.

• **Phosphorylation**: addition of a phosphate group to a molecule, often to modulate the activity of a protein.

• **Metal ion**: A metal atom that has lost one or more electrons, thereby acquiring a positive charge.

• **Kinase**: Enzyme that adds a phosphate group to a molecule, often to regulate the activity of a protein.

• **Ion channel**: membrane protein which allows the selective passage of ions across the cell membrane.

• **Cofactor**: Molecule which helps an enzyme to catalyse a biochemical reaction.

• **Retrocontrol**: Mechanism by which an end product of a signalling pathway inhibits an earlier step in that pathway.

• **Membrane receptor**: protein located on the cell membrane which interacts with ligands to initiate a cellular response.

• **Ionotropic**: Type of receptor that forms an ion channel and allows the passage of ions across the membrane in response to a ligand.

• **Metabotropic**: Type of receptor that activates intracellular signalling pathways without forming a direct ion channel.

• **Synaptic plasticity**: Capacity of synapses to strengthen or weaken their transmission in response to activity.

• **Signalling cascade**: sequence of biochemical events triggered by the activation of a receptor, leading to a cellular response.

• **Dysregulation**: Disruption of the normal regulatory mechanism of a biological process.

• **Excitotoxicity**: Neuronal damage caused by excessive stimulation of neurons, often due to high levels of glutamate or Ca^{2+}.

• **Metabolic syndrome**: A group of conditions that increase the risk of heart disease, stroke and diabetes.

• **Cerebrospinal fluid**: The fluid that surrounds the brain and spinal cord.

• **Dopaminergic neuron**: Neuron which uses dopamine as its main neurotransmitter, often affected in Parkinson's disease.

• **Chelator**: Substance that binds to a metal ion, facilitating its elimination from the body.

• **Signalling ionic** : Process by which the ions influence cellular responses to various stimuli.

• **Nanomedicine**: Use of nanoparticles to diagnose and treat diseases, particularly cancer.

• **Apoptosis**: Process of programmed cell death, essential for tissue development and homeostasis.

• **Gene therapy**: a technique that involves modifying the genetic material of a cell to treat a disease.

REFERENCES

- Alberts, B., Johnson, A., Lewis, J., Raff, M., Roberts, K., & Walter, P. (2015). Molecular Biology of the Cell (6th ed.). Garland Science.
- Barbagallo, M., & Dominguez, L. J. (2010). Magnesium and aging: a relationship with potential therapeutic implications. Current Pharmaceutical Design, 16(7), 832-839. https://doi.org/10.2174/138920110790442121.

- Berridge, M. J. (2012). Calcium signaling and cell proliferation. Cell, 149(6), 1177-1190. https://doi.org/10.1016/j.cell.2012.05.034.

- Berridge, M.J. (2016).Calcium signaling and cell proliferation. Biochemical Society Transactions, 44(5), 1141-1150. https://doi.org/10.1042/BST20160100.
- Clapham, D. E. (2007). Calcium signaling. Cell, 131(6), 1047-1058. https://doi.org/10.1016/j.cell.2007.11.028.
- D'Amico, M., & Magro, G. (2020). Metal ions as therapeutic targets in human disease. Nature   Reviews Chemistry, 4(10), 1-20. https://doi.org/10.1038/s41570-020-00267-2.

- Dyer, M. A., & O'Brien, D. (2017). In vivo imaging of metal ions in biological systems. Nature Reviews Chemistry, 1(8), 1-15. https://doi.org/10.1038/s41570-017-0073-2.

- Dittmer, P. J., & Kuhlmann, J. (2019). Techniques for the analysis of metal ions in biological systems. Analytical Chemistry, 91(3), 1805-1820. https://doi.org/10.1021/acs.analchem.8b04972.
- Gunter, T. E., & Gunter, K. K. (2001). Transport of calcium and

other ions in mitochondria. Biochimica et Biophysica Acta (BBA) - Bioenergetics, 1504(1), 1-27. https://doi.org/10.1016/S0005-2728(00)00200-5.

• Kambe, T., Tsuji, T., & Nagano, K. (2015). Current understanding of zinc homeostasis and zinc signaling in health and disease. Cellular and Molecular Life Sciences, 72(17), 3173-3196.https://doi.org/10.1007/s00018-015-1972-7.

• Kaczmarek, J. S., & Wysocki, M. (2021). Nanomedicine approaches for targeting metal ions in cancer therapy. Nanomedicine: Nanotechnology, Biology, and Medicine, 17, 1-15. https://doi.org/10.1016/j.nano.2020.102053.

• Koyama, A., & Hattori, N. (2015). Magnesium and Parkinson's disease. Current Medicinal Chemistry, 22(34), 4067-4074. https://doi.org/10.2174/0929867322666150902115145.

A. Krezel, A., & Maret, W. (2016). The functions of "metal-binding" proteins in the regulation of cellular zinc homeostasis. Journal of Biological Inorganic Chemistry, 21(5), 765-779. https://doi.org/10.1007/s00775-016-1382-5. Lodish, H., Berk, A., Kaiser, C. A., Krieger, M., Scott, M. P., & Bretscher,(2016). Molecular Cell Biology (8th ed.). W.H. Freeman.

• Mattson, M. P., & Camandola, S. (2018). Energy intake, meal frequency, and health: a neurobiological perspective. Annual Review of Nutrition, 38, 1-20. https://doi.org/10.1146/annurev-nutr-082117-051540.

• Pappalardo, G., & Sgambato, A. (2019). Signaling pathways in cancer: A review. Journal of Cancer Research and Clinical Oncology,

145(1), 1-12. https://doi.org/10.1007/s00432-018-2755-5.

• Rhyu, M. S., & Kim, D. H. (2019). The role of zinc in cell signaling and its implications in health and disease. Journal of Nutritional Biochemistry, 65, 1-10. https://doi.org/10.1016/j.jnutbio.2018.11.014.

• Rude, R. K. (2012). Magnesium deficiency: a cause of heterogeneous disease in humans. Journal of the American College of Nutrition, 31(2), 143S-150S. https://doi.org/10.1080/07315724.2012.10720269.

• Schuchardt, J. P., & Hahn, A. (2017). Magnesium and vitamin D: a systematic review. Nutrients, 9(5), 430.https://doi.org/10.3390/nu9050430.

• Takeda, A. (2003). Zinc and Alzheimer's disease. Journal of Nutritional Biochemistry, 14(6), 337-344. https://doi.org/10.1016/S0955-2863(03)00028-9.

• Zhang, Y., & Liu, Y. (2020). Advances in imaging techniques for studying metal ions in cells. Frontiers in Chemistry, 8, 1-12. https://doi.org/10.3389/fchem.2020.00001.

Printed by Books on Demand GmbH, Norderstedt / Germany